Threads books in hardback

Beans	Plastics
Bread	Rice
Bricks	Rocks
Clay	Salt
Cotton	Silk
Eggs	Spices
Fruit	Tea
Glass	Water
Milk	Wood
Paper	Wool

First paperback edition 1991

First published 1987 in hardback by
A & C Black (Publishers) Limited
35 Bedford Row, London WC1R 4JH

ISBN 0–7136–3502–9

A CIP catalogue record for this book
is available from the British Library.

Acknowledgements
Illustrations by Caroline Ewen
Photographs by Ed Barber, except for p 9, p 15 (top) Forestry Commission,
Edinburgh.

The author and the publisher would like to thank the following people whose
help and cooperation made this book possible: Caroline Pontefract and the staff
and pupils at Hampden Gurney School; Bert Wallis and the staff at Bowaters
UK Papermill, Kemsley, Kent; Dick Kemp, Scott Ltd; Dr R Cowling; Marion Jahan.

Typeset by August Filmsetting, Haydock, St Helens
Printed in Belgium by Proost International Book Production

Paper

Annabelle Dixon

Photographs by Ed Barber

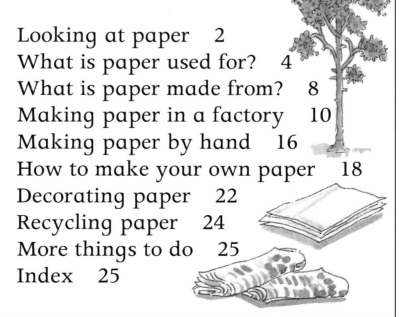

Contents

Looking at paper   2
What is paper used for?   4
What is paper made from?   8
Making paper in a factory   10
Making paper by hand   16
How to make your own paper   18
Decorating paper   22
Recycling paper   24
More things to do   25
Index   25

A&C Black · London

Looking at paper

Before you read this book, there is something you must do. It will not take long and it will help you to understand how paper is made.

Collect as many different kinds of paper as you can find. Here are some ideas . . .

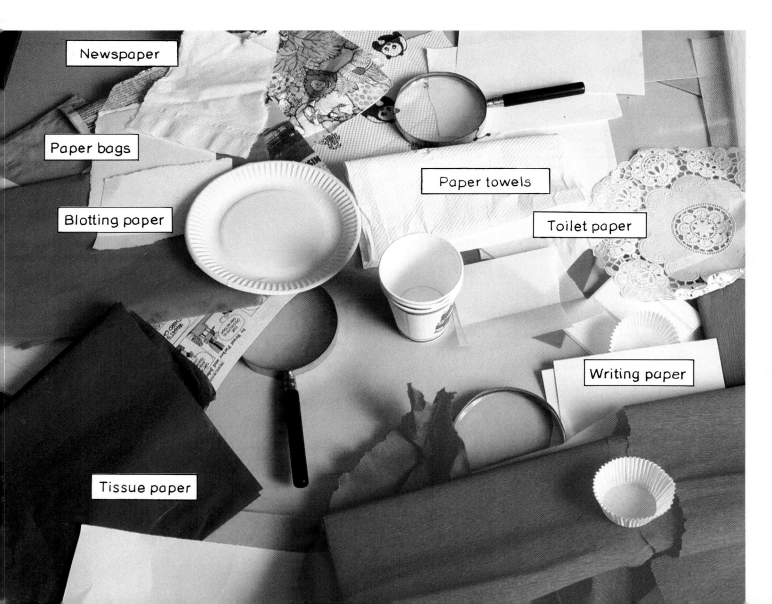

Newspaper

Paper bags

Blotting paper

Tissue paper

Paper towels

Toilet paper

Writing paper

Look really closely at the papers you have collected. Use a magnifying glass if you have one. How many differences can you find?

How easy is it to tear or fold each kind of paper? Scrunch up each kind of paper. What sort of noise does it make? Feel the papers with the tips of your fingers. Do they all feel the same?

What is paper used for?

You should find lots of differences between the papers you collect. This is because each kind of paper has a different job to do. To find out how paper is made, you need to look at these different jobs. Each kind of paper is made in rather a special way.

We use paper for all kinds of things.

▲ We make things with it.

We wrap things up in it. ▶

▲ We write, print and paint on it.

We clean and mop up with it. ▶

Paper for writing

When paper was first made, it was used mainly for writing.

Spread out all the papers that you collected at the beginning of the book. Find an ink pen or a felt-tip pen and try writing or drawing on each different kind of paper.

Are they all easy to write or draw on? Some papers, such as the paper towel or toilet paper, are made to drink up ink or other liquids. Other papers have a smooth, shiny 'skin' to stop ink soaking into the paper. This sort of paper is used for writing and printing. (Feel the corner of a page from this book.)

Making paper smooth and shiny

In a factory where paper is made on a big machine, substances such as china clay, a thin glue (called size) or plastics are added to make the paper smooth and shiny. The paper is also polished as it moves around very heavy metal rollers. These are called calender rollers.

In a few places, paper is still made by hand. The paper-makers polish the paper with a flint stone. This is called flint glazing. You can give a 'shine' to a piece of dull-looking paper by rubbing it with the back of a spoon.

What is paper made from?

Paper is made from long, thin threads called fibres. You can see the fibres if you tear a piece of paper and look closely at the torn edge. The fibres give paper its strength.

Thousands of years ago, the Egyptians used the fibres in papyrus stems to make paper. The word 'paper' comes from the name of the plant. But the paper they made was not very strong and did not last long.

The Chinese made the first real paper nearly 2000 years ago. They used old rags and rope and invented the way of making paper that we still use today.

Esparto grass

Spinach

Lettuce

Rhubarb

Hemp

Pine tree

Paper-makers went on using old rags and recycled material for hundreds of years. It made very good paper but so many people wanted paper that the supply of rags began to run out. Paper-makers then tried to make paper from different plants.

Wheat

Can you guess which of these plants would be best for making paper? *The answer is on page 25*.

Nettles

Paper from trees

The best plants for making paper contain strong fibres – like the stringy bits in runner beans. Trees are good for making paper because they have very long, strong fibres inside them. The most useful sort of trees are conifers, such as spruce or pine trees.

Conifer trees are specially grown on big plantations in countries such as Scandinavia, America and Indonesia. The trees are grown in straight lines so they are easier to look after and cut down.

Making paper in a factory

At the factory, the first step is to separate the fibres in the tree trunk from the stuff that holds the fibres together. This makes a substance called pulp. Pulp for poor quality paper, such as newspaper, and pulp for writing paper are made in different ways.

Pulp for writing paper is made by first cutting up the tree trunks into small pieces called wood chips – like the ones in this picture. Then the wood chips are cooked with lots of chemicals in enormous boilers as big as houses. This makes chemical pulp.

It costs a lot of money to make chemical pulp because at least half the tree gets wasted and the chemicals are expensive. The chemicals are also poisonous.

Pulp for newspaper is made by cutting up the tree trunks in a grinding machine with lots of hot water. This makes mechanical pulp. Mechanical pulp is full of little pieces and does not make very strong paper. But it does use up nearly all the tree and is cheap to make.

Some factories also make pulp by recycling waste paper. (Recyling means using something again.) Before it can be made into pulp, the waste paper has to be mixed with chemicals that take out the ink. The pulp made from waste paper is often mixed in with the new pulp made from tree trunks.

The pulp goes into large tanks called hydrapulpers, which are a bit like giant washing machines. Here the pulp is mixed with a lot of hot water and mashed into a thick, grey porridge, called stock. Paddles inside the hydrapulpers whisk up the fibres in the stock to make them fluffy. This helps the fibres to stick together.

At this stage, chemicals are sometimes added to make the paper shiny or coloured.

The stock from the hydrapulpers travels along pipes until it reaches a big machine, which turns the watery stock into dry paper.

The hydrapulpers and the machine for making paper may be linked to computers, which control the way the paper is made. The operator looks at the screens to check things such as how fast the paper is moving.

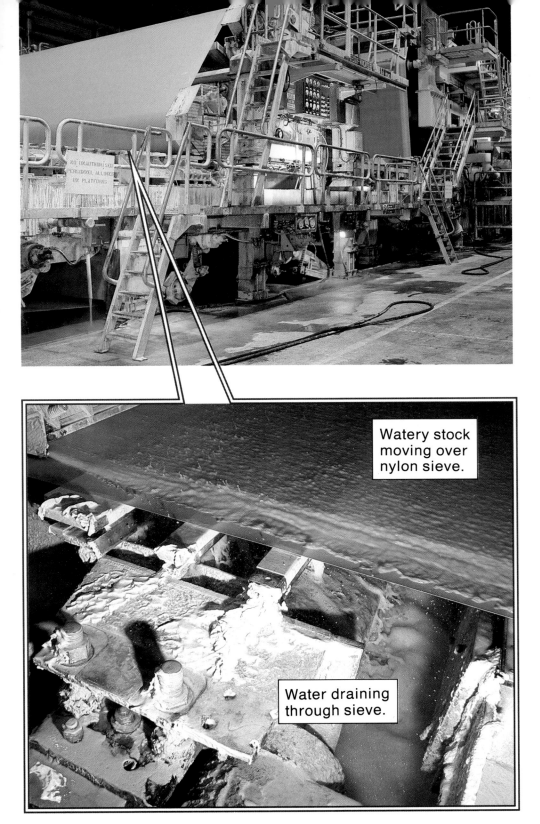

Watery stock moving over nylon sieve.

Water draining through sieve.

How does the machine work?

In a large factory, a machine for making paper is about as long as a soccer pitch and as wide as a house. The first part of the machine is a nylon sieve. The watery stock from the hydrapulpers moves at high speed over the sieve. As it moves along, some of the water drains down through the holes in the sieve.

Because the stock is nearly all water, this is called the 'wet end' of the machine.

At the end of the sieve, about $\frac{1}{4}$ of the water has drained through the sieve and a sheet of paper has started to form.

At the end of the sieve, the paper moves onto a thick felt pad. The felt helps to hold the paper together because it is very weak and could easily break apart. The paper moves around some large metal rollers, which squeeze out more water. The rollers also press the fibres together to make the paper the right thickness.

The paper then moves around another set of rollers, which are stacked one on top of the other in a long row. These rollers have steam inside to make them hot and the paper dries as it moves around them.

Sheet of paper

Very soon, a long, dry sheet of paper comes out of the other end of the machine. At this stage, the paper contains very little water so this is called the 'dry end' of the machine.

The paper is stored on huge rollers like a giant paper towel. The paper on each roll is long enough to stretch across the English Channel!

The rolls of paper are loaded onto lorries and taken to places such as printing factories, which use a lot of paper. Sometimes the paper is cut up into flat sheets before it leaves the factory.

Making paper by hand

Paper is still made by hand in some places today. This takes much longer than it does to make paper on a big machine, so hand-made paper is expensive. But it is good for painting because of the way the paper soaks up paint. Artists can also try out different ways of colouring and decorating the paper they make by hand.

1. This artist is making her own special sheet of paper. First she soaks pieces of white cotton in water.

2. Then she mashes up the wet cotton to make a mushy pulp and adds a dye to colour the pulp.

3. She puts the pulp into a bowl and stirs it well. Then she dips a wire sieve into the bowl and lifts out a thin layer of coloured pulp.

4. When a lot of water has drained through the sieve, she tips the pulp onto a damp cloth and presses it out of the sieve with a sponge.

5. She makes pieces of pulp in different shapes and colours and builds up a pattern on the cloth.

After leaving this to dry for at least 24 hours, she peels the finished piece of hand-made paper carefully off the cloth.

How to make your own paper

Would you like to try making your own paper? It is a good idea to wear an apron or some old clothes – you will probably get wet and messy!

You will need a sieve to drain the water out of the pulp. You can buy one from a craft shop – the proper name is a mould-and-deckle. Or you can make your own. Here is a good way to make your own mould-and-deckle.

1. Cut about 6 big holes in the lid of a square container of ice-cream or margarine.
2. Cut one leg off a pair of tights and stretch it carefully over the lid.
3. Make lots of holes in the tights with a sharp pencil.

To make your paper you will need

Plenty of old newspaper

A bucket half full of warm water

Clean blotting paper (or newspaper and paper towels)

Biological washing powder

An egg whisk or liquidiser*

An iron and ironing board*

A large, wide bowl

A rolling pin or an empty bottle

An old tablespoon

A clean piece of cloth

A sieve (mould-and-deckle)

To colour the paper you will also need

Powder paint or poster paint
2 tablespoons of paint for a light colour
4 to 6 tablespoons for a deep colour

*Ask an adult to help you with these things.

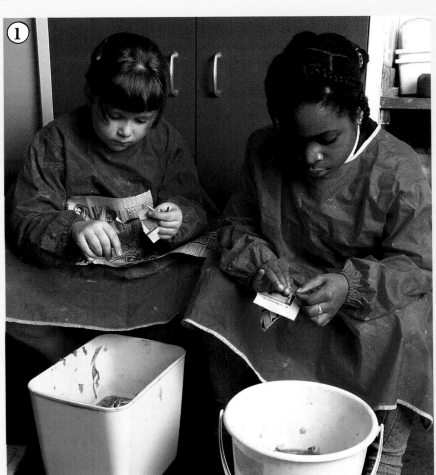

How to do it

1. Make the pulp first. Sprinkle a small amount of biological washing powder on to the warm water in the bucket. (Add colour now if you want to). Tear up several sheets of newspaper into very small pieces and drop them into the bucket. When the pulp is really soft (after 1 or 2 hours) tip out some of the spare water. Then scrunch up the paper with your fingers until the pieces are really small. Beat the mixture with a whisk (or use a liquidiser) to make a mushy pulp.
Turn over to find out what to do next.

2. Tip the pulp into the large, wide bowl and stir it well with the spoon. Next to the bowl, put a piece of blotting paper or a paper towel on top of a pile of newspaper.

3. Hold your mould-and-deckle upright and dip the edge into the bowl. Then slide it under the water and turn it so it is level with the bottom of the bowl.

4. Now carefully lift the mould-and-deckle up out of the water. You should catch a layer of pulp on the mould. Keep it flat and let some of the water drain back into the bowl. You can help by giving it a gentle shake. If the pulp looks too thick, tip it back and start again.

5. Turn the mould upside down and tip your pulp pancake quickly onto the blotting paper or paper towel.

6. Put another piece of blotting paper or another paper towel over the top of the pulp. Roll an empty bottle or rolling pin carefully over the top to squeeze out some more of the water.

7. Carefully turn your paper sandwich over and ask an adult to help you iron the bottom piece of blotting paper or paper towel. Put a clean paper towel or a clean cloth on the ironing board before you start ironing.

Turn your paper sandwich back again and carefully peel off the top piece of blotting paper or paper towel. Leave the paper to dry in a warm place. It will take one or two days to dry. If you would like your paper to have a smooth finish, leave it to dry on a smooth surface. When your paper is dry, peel off the bottom sheet of blotting paper or paper towel.

If you have not added any colour, the paper will be grey because of the printing ink in the newspaper. It will look like this. You can cut the edges to make them straight.

Decorating paper

Some paper-makers add decorations or coloured patterns to their paper. Would you like to try this?

You can add things to the pulp in the bowl.

Here are some ideas

Pieces of tin foil

Glitter

Seeds

Pieces of coloured paper or tissue

You could also try making the pulp from paper tissues or Christmas wrapping paper instead of newspaper.

Or you can lay things over the pulp after you have tipped it out of the mould-and-deckle.

Here are some ideas

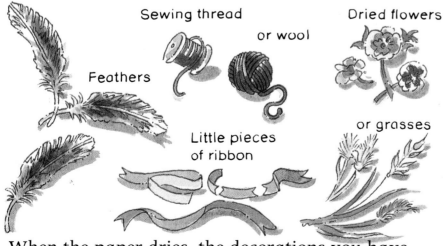

Sewing thread

or wool

Dried flowers

Feathers

Little pieces of ribbon

or grasses

When the paper dries, the decorations you have added will be part of your sheet of paper.

Marbling

You can decorate the plain paper made in a factory by making swirly patterns on the surface. This is called marbling because you can see the same sort of patterns in a stone called marble.

You will need

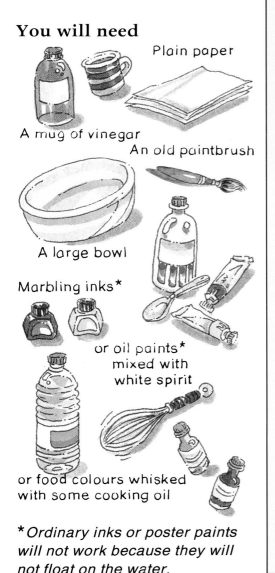

A mug of vinegar

Plain paper

An old paintbrush

A large bowl

Marbling inks*

or oil paints* mixed with white spirit

or food colours whisked with some cooking oil

*Ordinary inks or poster paints will not work because they will not float on the water.

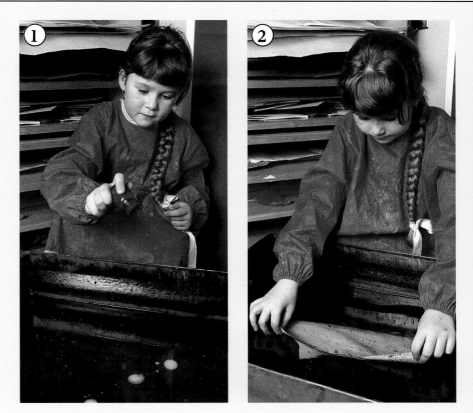

How to do it

1. Fill the bowl with water and add the vinegar. Choose about three colours and drop them carefully onto the water.

2. If you have a deep bowl, lower the paper slowly under the water. Move it slowly from side to side to make swirly patterns.

If you have a shallow bowl, lay the paper flat on the surface of the water. Wait until you see the patterns showing through and then lift up your paper carefully from one end.

You can clear the water of old colours by placing a piece of paper towel on the surface for a few minutes.

Recycling paper

Thousands and thousands of trees are cut down every day to make paper, especially newspaper. Too many forests all over the world are being cut down and not enough new trees are being planted.

You can help by saving your old paper and cardboard and giving it to a waste paper collector in your area. It can be recycled to make more paper, which will help to save trees.

More things to do

1. Try some of the tests paper-makers use to find out if their paper is well made.

The 'rattle' test. Hold a new piece of writing paper or brown paper in one hand and flick it with the fingers of your other hand. The sound you will hear is called a good 'rattle'. Then try the same test with an old paper bag. Can you hear the difference?

The 'bursting point' test. Collect some different kinds of paper and fold each piece in half. Then ask a friend to hold the corners of each piece of paper together while you drop weights, such as small stones, into the centre of the fold. When the paper tears, you have found the 'bursting point'. This test tells you how strong the different papers are.

2. Some papers have a thin covering, which is called a coating. Look back at page 7. The shiny surface on some kinds of paper is a coating.

Some coatings make greyish paper look brighter and whiter. Try rubbing toothpaste on one half of the grey paper you make from newspaper. Let the toothpaste dry and then iron it to make it smooth. Try writing on the toothpaste coating. Is it easier to write on than the paper without any coating?

Waxy or plastic coatings stop water, fats and oils from getting through the paper. Sweets are often wrapped in paper with a wax coating because it stops them getting sticky. You can make wax paper by rubbing a wax candle over a piece of writing paper. Then put a piece of clean paper over the top and iron it with a cool iron. (Ask an adult to help you with the iron.)

3. Greaseproof paper also stops things getting damp or sticky. It is made by soaking paper in special oils. You can make greaseproof paper by rubbing a little cooking oil on both sides of a piece of writing paper. Rub off any spare oil with a paper towel until it feels dry.

Page 8. Answer:
esparto grass, pine tree, hemp, wheat.
Try making paper from different kinds of plants.

Index

Numbers in **bold** type are pages which show activities.

blotting paper   2, **20**, **21**

calender rollers   7
Chinese   8
colouring paper   12, 16, 17, **19**, **23**
computers   12
conifer tree   9

decorating paper   **22**, **23**
drying paper   14, 15, **21**

Egyptians   8

factory,   7, 10–15
fibres   8, 9, 10, 12, 14

hand-made paper   16, 17
hydrapulpers   12, 13

machine   7, 12, 13, 14, 15
marbling   **23**
mould-and-deckle   **18**, **20**, **21**, **22**

newspaper   2, 10, 11, **19**, **20**, **21**, 24

painting   5, 16
paper towel   2, 5, **6**, **20**, **21**, **23**
papyrus   8
pine tree   8, 9
plantation   9
polishing paper   **7**
printing   5, 6, 15
pulp   10, 11, 12, 16, 17, 18, **19**, **20**, **21**, **22**

recycling   8, 11, 24

spruce tree   9

tissue paper   2, **22**
trees   8, 9, 10, 11, 24

waste paper   11, 24
wood chips   10
writing   5, **6**
writing paper   2, 5, **6**, 10